99 MORE
UNUSELESS
JAPANESE
INVENTIONS

Warning

What follows are simply suggestions, ideas, food for thought. Your mission, should you decide to accept it, will be to peruse the following pages at your leisure and treat your mind, spirit and imagination to a new and different form of 'unuseless' invention.

We don't suggest that you actually try to build any of these or similar gadgets, nor do we discourage it. We simply wish to caution you that Chindogu will not simplify your existence. At best they will distract from the other problems in life; at worst they will exhaust your time, money and storage space.

The art of unuseless invention is a double-edged sword. We take no responsibility for the turmoil, dissonance, obsession or other problems that may result. Proceed at your own risk.

About the Authors

Kenji Kawakami was part of the system back in the Sixties when he churned out story boards for an animated TV series. He broke free to arrange media events, become an avid opponent of karaoke, design the Tokyo Bicycle Museum and become editor of a popular home shopping magazine – in the pages of which Chindogu was born. He is the author of four Japanese-language books of unuseless inventions, founder and president of the International Chindogu Society, and the builder of hundreds of gadgets designed by himself and other members of the Society. Kenji currently works as an editor, photographer and part-time Chindoguist.

Dan Papia gave up a successful career as a financial journalist in 1991 to focus on more serious pursuits, and almost immediately created the so-called 'eye paddle' – a ball painted to look like an eye attached to an optician's chart paddle. He gave the Tokyo Dome its official name ('Big Egg'), wrote humorous columns for the *Mainichi* newspaper and the *Tokyo Journal*, and appeared irregularly for several years on Japanese TV. After becoming the world's first non-Japanese Chindoguist, he started the International Chindogu Society in LA. He also writes and consults for film and TV production.

99 MORE UNUSELESS JAPANESE INVENTIONS

The Art of *Chindogu*

Original photography, concept and
source material by Kenji Kawakami

Original and translated text by Dan Papia

W. W. NORTON & COMPANY
New York • London

The author is the Founder and President
of the International Chindogu Society

First American edition 1998

1 2 3 4 5 6 7 8 9 0

ISBN 0-393-31743-9

Printed in Italy

W. W. Norton & Company, Inc., 500 Fifth Avenue, New York, NY 10110

W. W. Norton & Company Ltd., 10 Coptic Street, London WC1A 1PU

Foreword

A message from Kenji Kawakami, founder of the Chindogu Universe

Have you seen the movie *2001: A Space Odyssey*?

In the first scene a hairy ape-like creature is standing at the water's edge in the middle of a wasteland. He picks up a bone, gets the idea of using it as a club and hurls it into the air. It flies higher and higher until it dissolves into a spaceship.

When we see scenes like this we realize that man's relationship with objects goes back twenty-five million years or so – a *very* long time – to the time of the cavemen. I imagine that that caveman (and he wasn't really even a man; let's call him a cave-ape) was probably history's first inventor. I also imagine that he was regarded as pretty unusual by his contemporaries. He probably discovered one day that a bone could be used for hitting and reaching for things that were out of his grasp. He may well have carried the bone around with him wherever he went. The others might have talked about him behind his back. But in his footsteps Edison and da Vinci would some day walk.

In his own time, each important inventor is different from his contemporaries. In Japanese we have an expression: as different as the moon and a snapping turtle. In the case of the cave-ape the difference was probably even more pronounced. He was as different as one of Mars's moons and a snapping turtle.

I introduced the term Chindogu in Japan in 1985. Coined from 'chin' meaning unusual and 'dogu' meaning tool, it refers to a most universal concept: a gadget that appears to be useful but on closer examination isn't. These inventions must exist, but they cannot serve a reasonable function. Inherent in the premise of Chindogu is this fundamental contradiction.

In creating the institution of Chindogu, I believe that I have come up with a new way of thinking that will give this cave-ape the respect he deserves. It is an unprecedented form of humorous culture that allows us to get back to the true joy of inventing. Since its introduction in Japan in 1985, some six hundred Chindogu have come into existence. When we look at them, we laugh, but we aren't mocking the inventor. We laugh because, for a fleeting moment, we almost think we can use the object. We nearly consider buying one. When we laugh, we are laughing at ourselves. The uniqueness of Chindogu comes from the fact that we later realize that in the course of applying our creativity and eliminating the old inconvenience we've simply caused a new – even bigger – inconvenience.

Let's play a little game. Pick up some of the 'essential' objects in your life – saucepans, CDs, *some* of your clothes – stash them in a cupboard, and see how you get by without them. Did your world become so inconvenient

that you couldn't stand it? No; you'll probably have found that you weren't inconvenienced at all by their absence. So what have you learned? The exercise gives you a chance to think about how these things have infiltrated our lives when we don't really need them. Chindogu point out to us that a life without our 'things' is a life of freedom. When we find ourselves sucked into the modern world's ever-swelling commercial culture, Chindogu provide an antithesis. They push against capi-

talism and give birth to a new form of art. They revitalise our human spirit and consciousness by steering clear of the trappings of fame and fortune. Chindogu have taken a different path from inventions built and patented to make money. They reside on a higher plane, removed from greed, where concepts like purity, innocence and nobility reign. The Art of Chindogu is truly proud and truly regal.

Since their introduction to the world, Chindogu have popped up all across the planet. They've appeared in movies and on TV. They've been in museums and trade shows. They've spiced up books and magazines. They've carried newspaper columns and radio shows. They've even established a presence on the internet. Chindogu has taken the world as its stage and continues to grow in popularity.

Ladies and gentlemen worldwide, I hope that you will join me in clearing your heads, opening your hearts and enjoying the ninety-nine more 'unusual tools' that have been selected for you to peruse and ponder.

Kenji Kawakami
July 1997
Tokyo, Japan

Chin-doh-what?

珍道具

Relax. Chindogu is not some ancient Japanese art form analogous to flower arranging or a tea ceremony. Nowhere in Asia – or in the world for that matter – are archaeologists ever likely to uncover the remains of a Back Scratcher's T-Shirt or a Rotating Spaghetti Fork while unearthing prehistoric pots and spearheads.

A Chindogu, or unuseless invention, is a gadget that has been conceived, designed, built, tested and verified to make our lives more convenient in some way. The only problem is that a Chindogu also makes our lives more inconvenient in another way. The Tongue Cover, for example, prevents us from burning ourselves when we drink a cup of unusually hot coffee, but it also prevents us from tasting it. Our Cranium-grabbing Glasses pleasantly take the weight of our spectacles from our ears, but they put it all on our forehead.

Coming up with gadgets that will qualify as Chindogu takes a little patience, a little practice and a lot of expendable time. But we've tried to make it as simple as possible for the would-be Chindoguist to ascend to the state of unuselessness and grow that third eye in the back of the head. If you adhere rigidly to the Ten Tenets of Chindogu as you tinker – discarding everything that either doesn't work at all or does work but works too well – eventually you should succeed.

In the ninety-nine gadgets that follow, the spirit of a whole new model for interplanetary communication resides, non-verbal, non-cultural, and non-useless. These particular inventions happen to come from Japan, but the heart of Chindogu is everywhere.

A gadget that works but doesn't work, helps but doesn't help, is useful but isn't . . . sad but funny, smart but stupid, logical but inconsistent . . . and at the same time pure, basic and universal. That's all we've really ever needed. And that's all Chindogu really is.

The Ten Tenets of Chindogu

Every Chindogu is an 'unuseless' object, but not every 'unuseless' object is a Chindogu. In order to transcend the realms of the merely unuseless and join the ranks of the really unuseless, certain criteria must be met. It is these criteria, a set of ten vital tenets, that define the gentle art and philosophy of Chindogu. Here they are:

1. A Chindogu cannot be for real use
It is fundamental to the spirit of Chindogu that inventions claiming Chindogu status must be, from a practical point of view (almost) completely useless; i.e. 'unuseless'. If you produce something that turns out to be so handy you use it all the time – everyone wants one and not a single person laughs – you've failed to make a Chindogu. Try the Patent Office.

2. A Chindogu must work
Ideas are good, but when it comes to Chindogu ideas alone aren't enough. You have to be able to use the object. It's not going to be useful, but it needs to be usable. Only after designing it, building it and testing it are you then qualified to decide that it wasn't worth the effort.

3. Inherent in every Chindogu is the spirit of anarchy
Chindogu are man-made objects that have broken free from the chains of usefulness. They represent freedom of thought and action: the freedom to challenge the suffocating historical dominance of conservative utility; the freedom to be (almost) useless.

4. Chindogu are tools for everyday life
Chindogu are a form of non-verbal communication understandable to everyone everywhere. Specialized or technical inventions, like a three-handled sprocket loosener for drainpipes centred between two under-the-sink cabinet doors – the unuselessness of which will only be appreciated by plumbers – do not count.

5. Chindogu are not for sale

Chindogu are not tradable commodities. If you accept money for one, you surrender your purity. They must not be sold, even as a joke.

6. Humour must not be the sole reason for creating Chindogu

The creation of Chindogu is fundamentally a problem-solving activity. Humour is simply the by-product of finding an elaborate or unconventional solution to a problem that may not have been that pressing to begin with.

7. Chindogu are not propaganda

Chindogu are innocent. They are made to be used, even though they cannot be used. They should not be created as a perverse or ironic commentary on the sorry state of mankind.

8. Chindogu are never taboo

The Chindogu Academy has established certain standards of social decency. Cheap sexual innuendo, humour of a vulgar nature and sick or cruel jokes that debase the sanctity of living things are not allowed.

9. Chindogu cannot be patented

Chindogu are offerings to the rest of the world – they are not ideas to be copyrighted, patented, collected and owned. As they say in Spain, mi Chindogu es tu Chindogu.

10. Chindogu are without prejudice

Chindogu must never favour one race or religion over another. Young and old, male and female, rich and poor – all should have a free and equal chance to enjoy Chindogu.

珍道具

Stormy Sky Tie
✳ *A must-have garment for wicked weather*

When you've got to lug an umbrella around just because the weathermen predict the chance of showers, you feel like you're getting the short end of a hook-tipped stick. And, if it doesn't rain, you'll probably leave it in a taxi.

The solution: an umbrella around the collar in the form of a tie. If you need it, it's there. And if you don't, you're still getting some use out of it. As a bonus, the added weight stops the tie blowing over your shoulder in a storm and it's lighter to carry than it would be at hand level.

Only one small Chindogu snag remains. When the umbrella is put into action, it makes for a wet shirt and suit when it reverts back to being a tie. But don't worry! If you never had the umbrella in the first place you'd have got wet anyway.

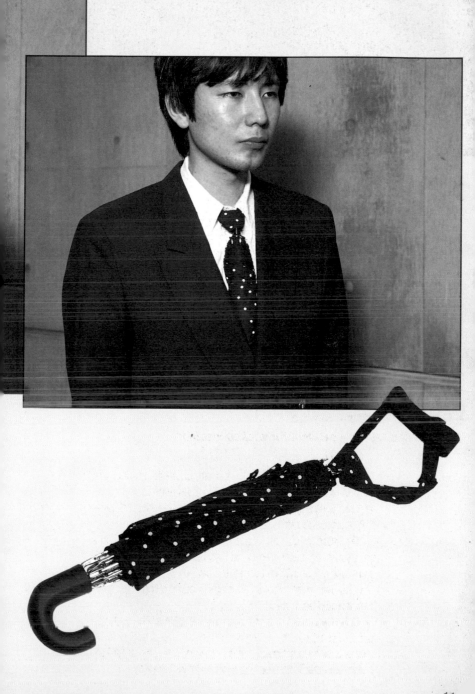

Portable Armrests
✳ *With these in place, life's one big armchair*

At the end of a hard day, luxury is a cushy recliner with padded armrests. Dining-chair uprights just don't give the same relief. So if fifty percent of that easy feeling is down to laid back arms, give your biceps a break even when walking. With Portable Armrests walking has never felt so good. While your legs take the strain, and your shoulders complain, your arms will be in paradise.

Rotating Spaghetti Fork
✳ *Takes the nuisance out of noodle nibbling*

The inventor of the tradition of eating spaghetti with a fork must have been a real practical joker. How is it possible to eat spaghetti and retain your dignity? At long last, the Rotating Spaghetti Fork is the answer.

With two small batteries and a motor that doesn't spin too fast in your utensil, you can turn pasta consumption into a smoother and more enjoyable enterprise.

Seeing Eye Seals
✳ Give the appearance of paying attention

Want to sleep through lectures? Want to look alert rather than inebriated after lunch? Drooping eyelids giving the game away? From any teacher's or employer's point of view, you'll always be the most bright-eyed and bushy-tailed person in the room with Seeing Eye Seals. In fact, your educators and employers would probably love to have some of these themselves.

And, just think, on camping trips tigers won't sneak into your tent if they think you're awake. A pair of these could ward off everything and everyone – pests, pickpockets and policemen. We just don't recommend driving in them.

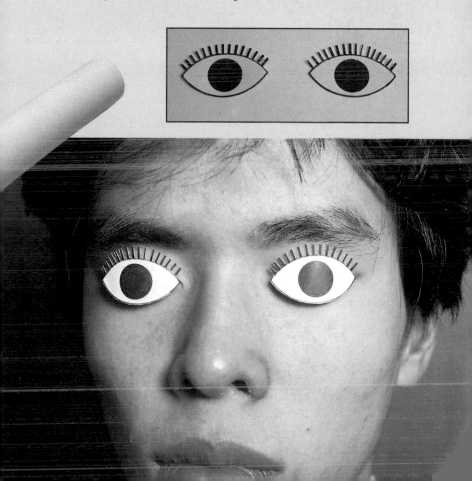

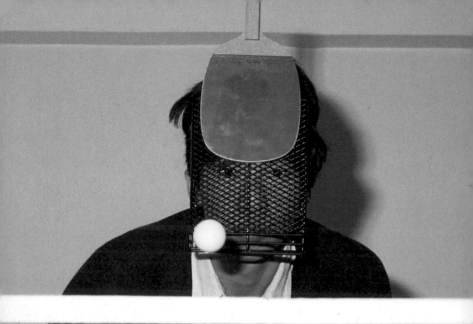

Face Ping Pong
✳ Give your head a workout

The challenge of designing a new international sport presented itself to us and, rising to the occasion, we came up with Face Ping Pong. The rules are simple. Keep your hands at your sides and bring the racket to the ball using only your head. Like most popular sports, the participants get to enjoy stimulating activity which they don't normally have the chance to indulge in, i.e. hitting things with your face. You'll find it every bit as fun and challenging as football or tennis and you will be sweat-soaked in next to no time. It's great exercise, but be prepared for a stiff neck the day after.

Last Bite Bar

✳ *The elegant way to clear your plate*

How many times could you have got away with using just a single fork or spoon were it not for that elusive last bite? When almost all your food is eaten and there's nothing left to use as leverage, you either need more silverware or to create a very smart diversion. The limitations of traditional cutlery mean that all too often diners are forced to submit their plates to waiters, bidding tearful farewells to those missed nibbles.

Those bites add up, so take what's rightfully yours and put hunger on hold. The Last Bite Bar can deliver that final dollop to your mouth.

Gift of the Gab Glasses
✳ *A social leg up for the strong silent type*

Some people just aren't good at talking, and the problem is that, despite the technological revolution, speech is still the most popular form of communication.

If you're shy or silent, don't put yourself at a disadvantage. Our Gift of the Gab Glasses enable even the most tongue-tied talker to give his audience the impression he's said at least three times as much as he actually has. But that's not all. They also render the listener speechless, making the wearer's conversation seem all the more skilled.

Walking around in these can be difficult, and glancing into a mirror downright scary. But it's good to talk.

珍道具

Shin Shines

✴ **Give shoes a glow, even on the go**

Shiny footwear speaks volumes. But the rub is that when you're an accomplished businessman there simply isn't time to rub your shoes. Shin Shines iron out this irony. At the bus stop or in the boardroom, brush-brush, wipe-wipe, and your shoes are again ready to attest to your success.

For the time being, ladies will just have to rely on the old trick of having closets full of extra shoes.

Extra Step Slippers

✳ *A ten-centimetre lift – at the drop of a cup*

When tall people need to be shorter, they can stoop. But unfortunately this simple theory doesn't work in reverse. The smaller members of our species are constantly having to appropriate ladders in order to reach things that are tantalizingly out of reach.

Get even with Extra Step Slippers. These light-weight booster cups are always poised to snap into place when an extra lift is needed. Walking in them may be uncomfortable, but so is a permanent stoop.

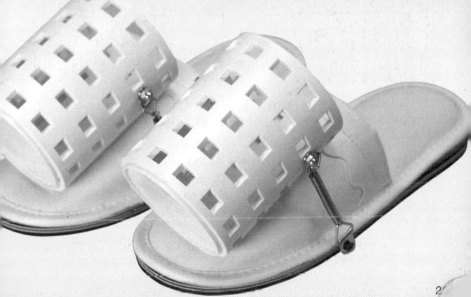

Four-for-One Floral Enhancer

✴ Saves money and space and looks great!

Longing to fill your home with the fancy floral arrangements found in glossy home decorating magazines? Strapped for cash? Dubious about the longevity of your blooms and unhappy with artificial alternatives? Make your flowers worthwhile. Three roses make a dozen when the Floral Enhancer is in place.

Business Tool Belt
✳ *Everything you need for the office – in one place*

Telephone repairmen and electricians get to carry all their important tools around their waist and are always poised for action. The average businessman has to make do with jotting down a note whenever he gets the urge, then waiting until he's back at the office. Not any more. With the Business Tool Belt name cards, telephone, calculator, bank slips and address book can be kept close at all times. The Business Tool Belt can get a little heavy, but maybe that's how repairmen stay in such good shape.

Wheels Wherever

✳ Removable rollers get things going

Some things are so heavy the manufacturers build them on rollers: filing cabinets, fridges, suitcases – objects we wouldn't be able to move any other way. Now see what wonders wheels can do for all those lighter objects too.

With adhesive-topped roller sets, just peel the seal and add a wheel. Make your household fully mobile. 'Where's the dictionary?', 'Could I have the mustard?' – just roll it over or pass it with a push. A word of warning: try not to think about what might happen in an earthquake.

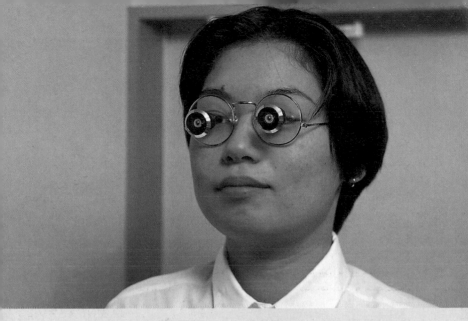

Wide Angle Glasses
✳ *Make tiny apartments into castles*

Sky-high rents in large cities make it impossible for the average person to have more than a room or two. The estate agent's spiel about 'One Room Mansions' sounds appealing, but your flat will still have about as much space as a rabbit hutch.

That's where Wide Angle Glasses come into their own. Like looking through a front door peep hole, with these on, the other side of a ten-foot room seems simply miles away.

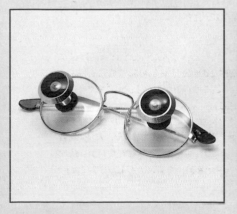

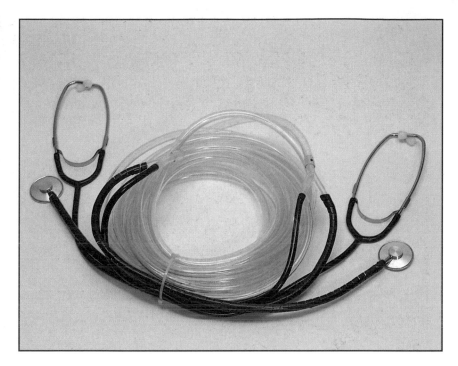

Battery Free Mobile Phone

✳ ***No charging, no phone bills, no worries***

It's easy to forget that the simplest solutions are always the best. Private calls on cellular waves are often hacked into by eavesdroppers; important information is at risk. Avoid all this hassle by saying no to portable phones.

With twenty feet of hose, the Battery Free Mobile Phone has plenty of scope to allow cost-free communication between two apartments in the same building, even two cars on the same road. There are no batteries, so you don't have to charge it up and there's no static, so you don't have to yell.

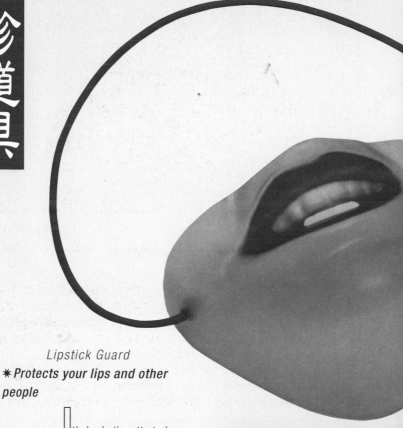

珍道具

Lipstick Guard
✳ Protects your lips and other people

It's bad etiquette to leave lipstick marks on people, especially those you don't know. But in crowded trains and cramped lifts, it can be hard to avoid close encounters. Try the Lipstick Guard for size.

With a discreet air hole to allow breathing, speaking, even smoking, the Lipstick Guard will not only protect the lips of the wearer, it also allows you secretly to chew gum or sleep with your mouth open without causing offence (provided you don't snore, of course).

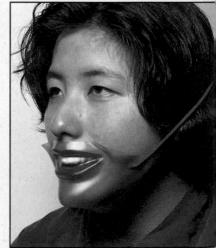

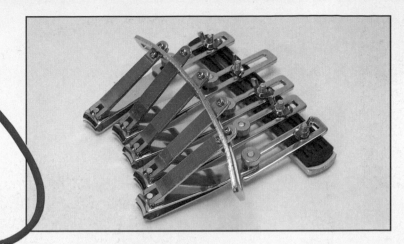

One Cut Clippers

✳ Don't like cutting your nails? One snip and it's all over

If you have an intolerance for the menial chores of personal grooming, this is the perfect short-cut for you. One Cut Clippers trim all the nails on a single hand in one fell swoop, with the added perk of cutting them all the same length. Flip it over and another clip tidies up the other hand. Easily adjustable, one set lasts a life time.

But hold your hand too far one way and you'll miss a nail. Hold it too far the other and you'll get part of your finger too. And don't forget: if you like having all your digits, adjust to the exact size of your hand or foot before use.

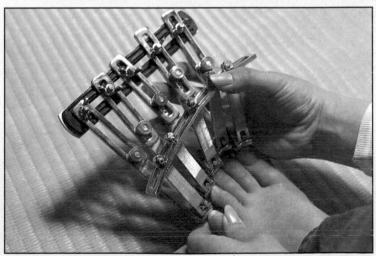

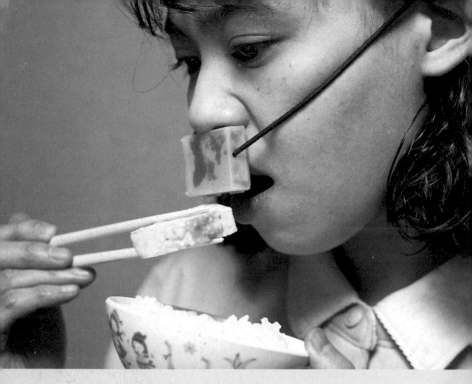

Aroma Eater

✳ ***Dieting's a snap when everything smells so good***

The problem with healthy eating is that nutritious, good-for-the-body food just isn't exciting.

Since when did overweight people, struggling to take off the pounds, crave low-calorie food? Where's the fun in a lettuce leaf and a poached egg? After day three of your diet you never want to see another green bean again.

Working on the simple principle that smell accounts for a large proportion of taste, the Aroma Eater is the answer to the dieting dilemma. For the same reason that cup cakes don't taste nearly as good when eaten next to paper mills, bland, healthy foods are likely to go down a lot better when the eater is bombarded with the aromas of cocoa, cinnamon, bacon grease and parmesan cheese.

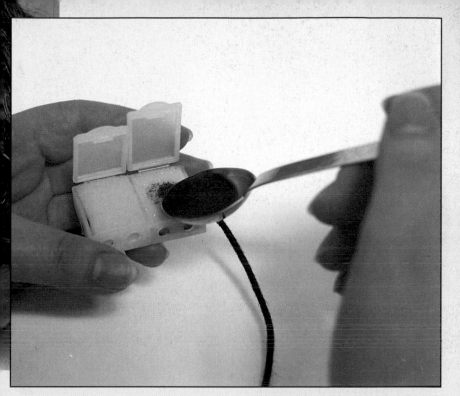

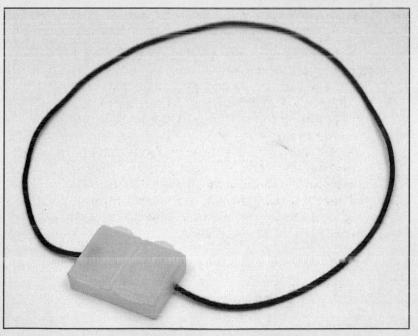

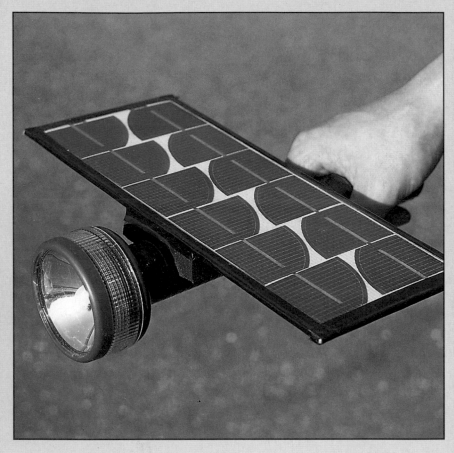

Solar Powered Flashlight

✳ **For the visionary environmentalist**

Dispense with batteries. Take one solar panel and a torch and see how easy it is to harness light from the sun, converting it into six volts of pure power and a bright beam.

Brilliant, economical, effective, innovative. This archetypal Chindogu would be so completely useful were it not so completely useless.

Cranium Camera
✳ Keep ahead of the crowd

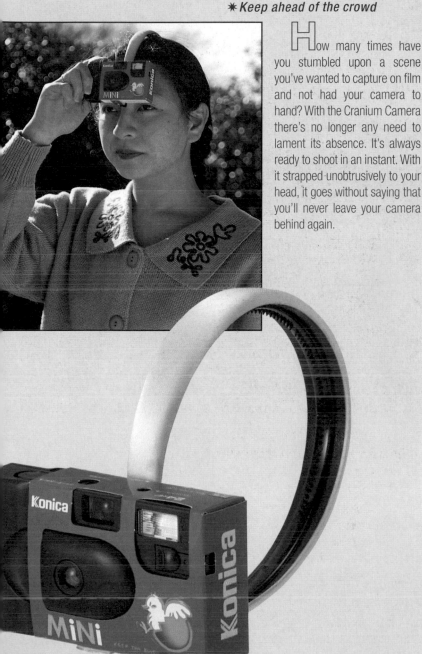

How many times have you stumbled upon a scene you've wanted to capture on film and not had your camera to hand? With the Cranium Camera there's no longer any need to lament its absence. It's always ready to shoot in an instant. With it strapped unobtrusively to your head, it goes without saying that you'll never leave your camera behind again.

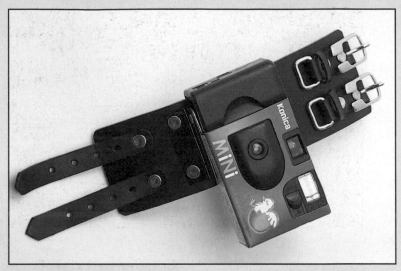

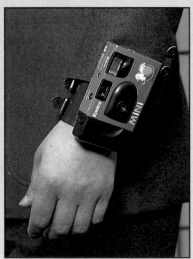

Wrist Camera

✳ When you need a camera...and quick

        Consider the wristwatch. If you can wear a clock on your wrist, then logic asks why not a camera? The Wrist Camera keeps your hands free and your camera on standby. At the drop of a hat, and the flick of a wrist, you can snap the perfect snap. If you're a spy, or maybe a paparazzi photographer and discretion is necessary, you can simply slip your hand in your pocket.

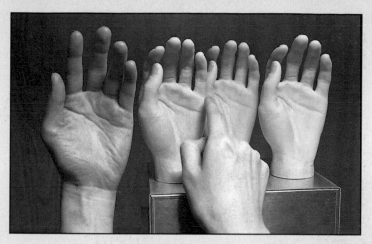

Extra Fingers

✳ If maths isn't your strong point, you can count on these

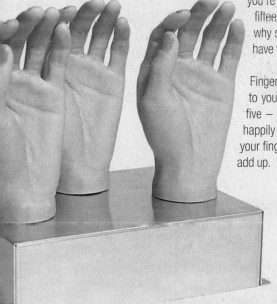

Mental mathematical blocks are a serious business. A lack of numerical prowess may result from a pathological dislike of your maths teacher, or from a debilitating crush on her. Alternatively, you may be a compulsive doubter and no matter how many times you're assured that six plus nine is fifteen, you just can't be sure. And why should you be when you only have ten fingers to count on?

Each block of Extra Fingers adds another fifteen digits to your arsenal (five plus five plus five – trust us). This way you can happily calculate small amounts on your fingers and physically see them add up.

Everywhere Exerciser
✳ Sitting, walking, standing, lying, eating...bench pressing

Allegedly, being a sports - man is a twenty-four-hour job. The same is true of a Chindoguist. So, luckily for athletes the world over, the Everywhere Exerciser is an indispensable workout machine for the bathroom, bedroom, work-room, waiting room, conference room, and even for destinations in between.

For business people who don't get enough exercise, including bank clerks looking to improve their figures, the Everywhere Exerciser is a workout tool that straps around your chest so that you can never 'accidentally' leave it somewhere. You'll be exercising in restaurants, trains and on the street. Now, unfortunately, there's no excuse.

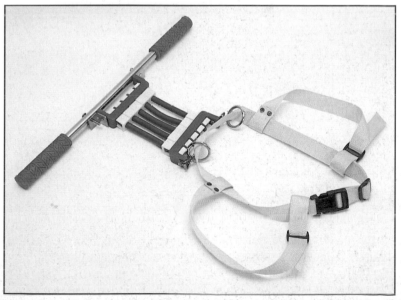

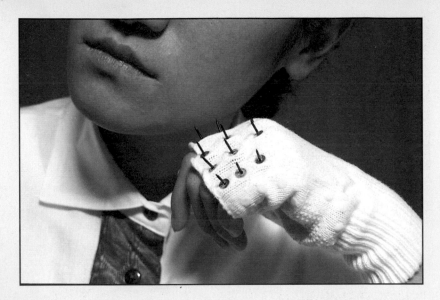

Lean Prevention Gloves

※ ***Gloves that make sure you keep your chin up***

Sit at a table and before you know it you'll be tempted to rest your head on your hand. The next step is leaning over, falling asleep, or ruining your posture so that you wind up with a sore back and a stiff neck.

Lean Prevention Gloves solve all of the above by nipping the problem in the bud. If you slip up and let your impulse to lean get the better of you, chances are it won't happen again. It's been tested.

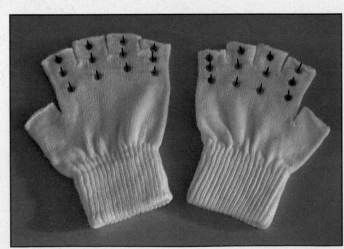

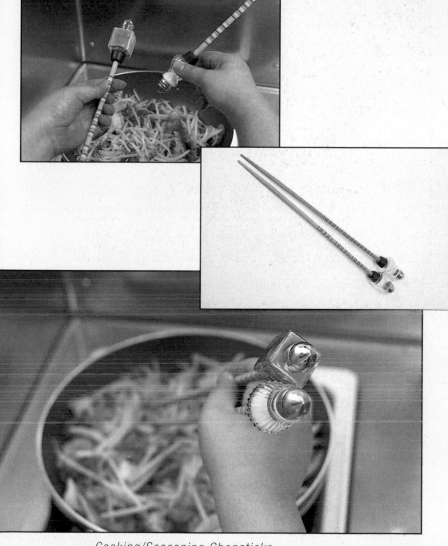

Cooking/Seasoning Chopsticks
✳ *Do more for your dishes with less effort*

Stir-fry cooking is quick and easy. One fire, one pot and, if you're eating alone, no dishes. Could anything be simpler? Yes. Now you don't even have to reach for the salt and pepper. They're at the end of your chopsticks! And if the food isn't spicy enough for your taste, adding more is easy to do – even while you eat.

Nature Lovers' Footwear

✳ For natural soles

It's hard to be a naturalist and a working professional at the same time. You want to be lying nude against a tree, but your boss wants you at a desk in a three-piece suit.

By wearing Nature Lovers' Footwear, your shoes will allow you to compromise. Commune with nature with each step you take, earn that executive's salary, and keep your freedom as your little secret.

Watch out for broken glass and soggy chewing gum.

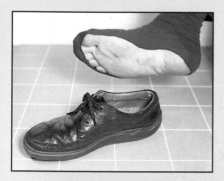

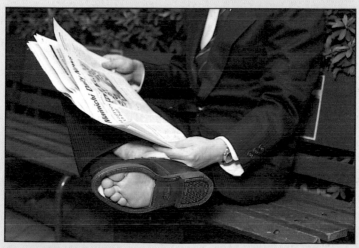

Everything Apron

✳ Want to be Queen of the Kitchen?

The art of good cooking is the delicate balance between art and science, and access to the right tools. Why settle for an apron with just one or two pockets on the front? The handier the tools, the better. Dress for action like any true technician. The Everything Apron brings the scouting notion of being prepared into the realm of gourmet.

Now you can cook more quickly and proficiently, with all you need just a hand's length away. And, with the added benefit of being the only one in the kitchen who can get to the tools, you'll never again suffer from too many cooks spoiling the broth.

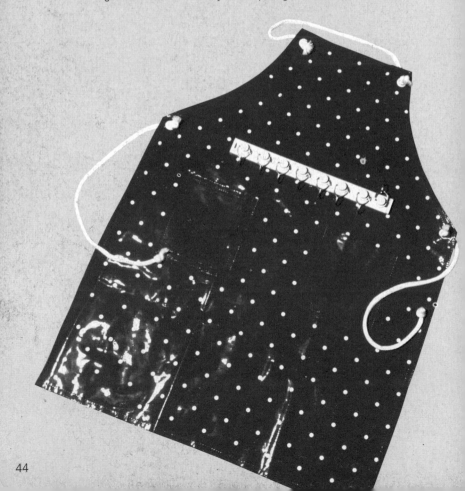

Paint-on Plate Protector
✳ *Takes the wet out of washing-up*

Lots of people love to cook, to lay their creations out on elegant china, then devour them. That's the fun part, but few enjoy washing up afterwards. With a bottle of Paint-on Plate Protector, you don't have to any more. If you've layered your plates before laying out your food, cleaning up is merely a matter of lifting the protector and dumping the remains. The plates are as good as new and ready to be used again.

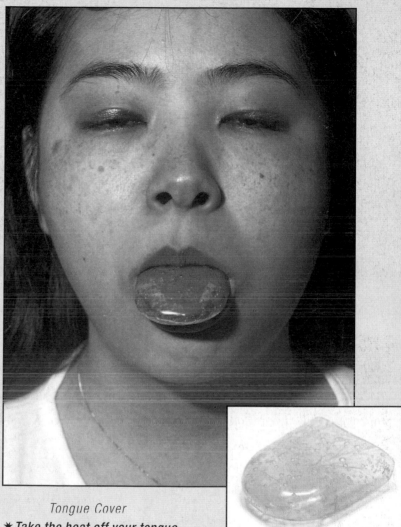

Tongue Cover

✳ Take the heat off your tongue

Tongues aren't always attractive but, boy, are they useful. They help us taste, talk and tease. But when you're really hungry and your food's really hot, they tend to get in the way. But not if you're wearing the Tongue Cover. Now you can shovel away fiery curry after steaming soup after boiling coffee and satisfy your stomach without harming that sensitive tongue. Possible drawback: you may find that your taste buds are a little less receptive.

1

maruman

QUARTZ

Personal Wake Up Headphones
✳ Rouses the deepest of sleepers...in stereo

Can't seem to get out of bed in the morning? No response to alarm clocks that beep, buzz or play music, snooze alarms that go off every few minutes, timers that turn on lights? When nothing else works, it's time to wake up, smell the coffee and get some Personal Wake Up Headphones.

The trick is that your personal headphones won't just nudge you into consciousness, they'll startle you into absolute panic. It's by no means a serene way to start the day, but with your heart fluttering and your nerves jangling there's no way you'll be dozing off again five minutes later.

But you must be a really sound sleeper to use these. If not, the ticking won't allow you to get to sleep in the first place!

珍道具

Dog Day Shoes

※ **Walk this way, and let the dog take the blame**

Wouldn't it be great if we could run as fast as dogs, we thought. Wouldn't it be even better if we had their ability to grip? But wouldn't it be best of all if we had shoes that left dog prints in the sand or in the mud rather than foot prints or shoe prints? Then we would be able to track through mud and onto carpets, over wet cement and onto glistening floors, and have everyone cursing some non-existent canine for the dirty deed. It's not a dog's life after all!

Diet Dishes
✳ *Satisfy your hunger with a culinary optical illusion*

We all know that the difficult thing about dieting is going without. Either you aren't able to eat the foods you want, or you have to eat less of them – either way you feel short changed. Not so much because your stomach gripes but because your brain isn't getting what it's used to.

Con it with Diet Dishes, which enable you to serve up full portions of your favourite foods while only consuming half the quantity. It's half as many calories, half the cost and half the dining time. Seems like magic? It's all done with mirrors.

Pet Slimming Bowls
✳ *Cuts calories without cries*

Cat too fat? Dog a hog? Monkey getting chunky? Pet Slimming Bowls, the animal version of the Diet Dishes, work just as well as their human counterparts. It's not often that a dog will acknowledge it has over-eaten but the tell-tale signs will soon become apparent.

If your clever canine knows he's been duped and howls until you give him another portion, don't fret – all that crying is bound to burn off some mutt gut.

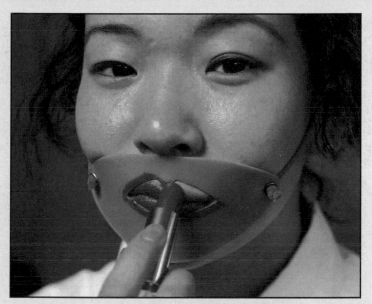

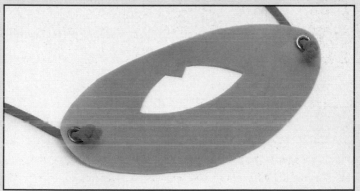

Smearless Lip Stencil

✳ *Kiss goodbye to smudged edges*

Rushed, riled and unrested, a busy woman needs all the help she can get making herself beautiful in a hurry. The Smearless Lip Stencil is a fool-proof way to get your lipstick where it belongs: that's not on your teeth or on your chin, but on your lips.

But make sure your stencil is custom made. What makes one lovely lady beautiful could leave another wishing she'd opted for the natural look.

Kissing Coffee Cup

✳ **With such a sweet cup there's no need for sugar**

If you're not getting a big enough kick from your coffee, make it more stimulating – and more fun – with this turn-on of a coffee cup. After all, people crowd into coffee clubs not simply because the black liquid they pour is so much better than the home-made version – it's because of the atmosphere. So when a spoonful of instant and some hot water alone don't look like they're going to do it for you, try drinking your coffee from a Kissing Coffee Cup. It makes your coffee break a much sweeter occasion.

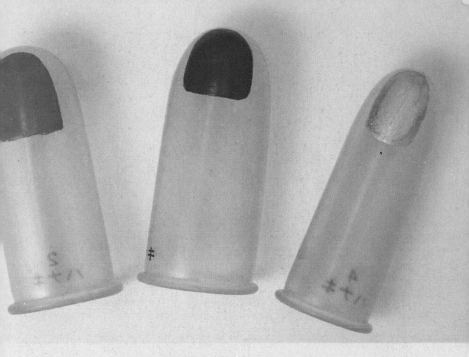

Pre-coloured Nails
✳ Manicured fingernails the way you want them – in an instant

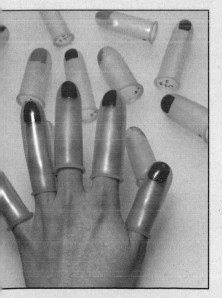

Ladies, it can take ages to file, polish and dry your nails. You get them the way you want them, go out dancing for the evening and come home only to realize you'll be wearing a different colour outfit the next day. Time to start all over again!

But not if you have a few sets of Pre-coloured Nails which can be switched at a moment's notice. No mess, no polish, no continual expense, and they're small enough for extra sets to be taken along in a handbag   just in case. The only quicker way to change your nail colour is to stick your fingers in a bucket of paint.

Backward Print Pads
✸ Shoes with soles that keep your secret

      Shoes are a dead give-away about who we are and where we're heading. This basic flaw in footwear design can literally be turned to the advantage of the savvy spy, or wary walker, with Backward Print Pads. These second soles that attach to the bottoms of your shoes send would-be assailants in completely the wrong direction.

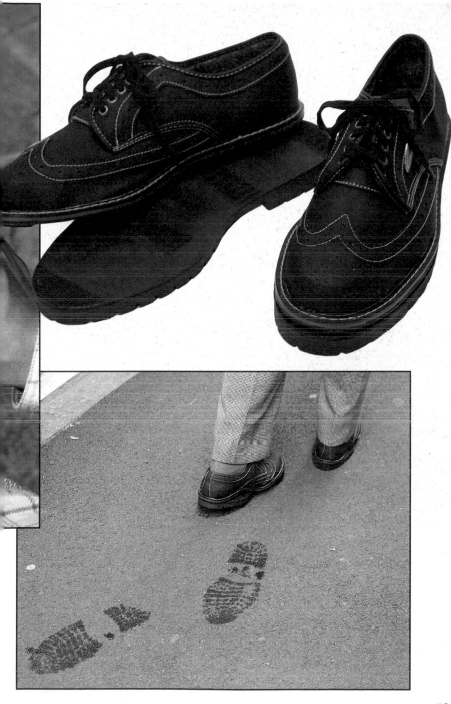

珍道具

Near Far Glasses

✳ See what's up close and way in the distance...at the same time

The telescope/microscope lenses of Near Far Glasses allow the wearer simultaneously to see both fine detail and objects on the horizon, eradicating the need to shift focus. You can read the fine print in your book while you wait for a taxi and scout for taxis streets away at the same time. Without adjusting your gaze, you can see the tiny markings on your map and check them against the landmarks that surround you.

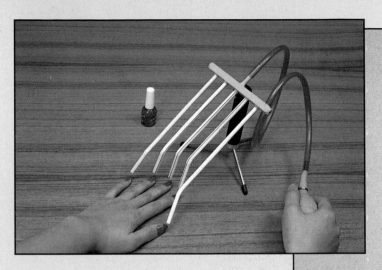

Polish Pump

✳ *For dry nails – quick*

Waiting for nails to dry seems to take forever. And, while you're waiting, if you accidentally brush them against something they'll have to be redone. Complete immobility isn't much fun.

To speed things up use the Polish Pump. Squeeze the air ball with one hand and the five tubes each direct a stream of air directly onto the five fingernails of your other hand.

You might ask, why not simply connect the tubes to an electric fan? Well, this way, one of your hands always has something to do. The time will just whizz by.

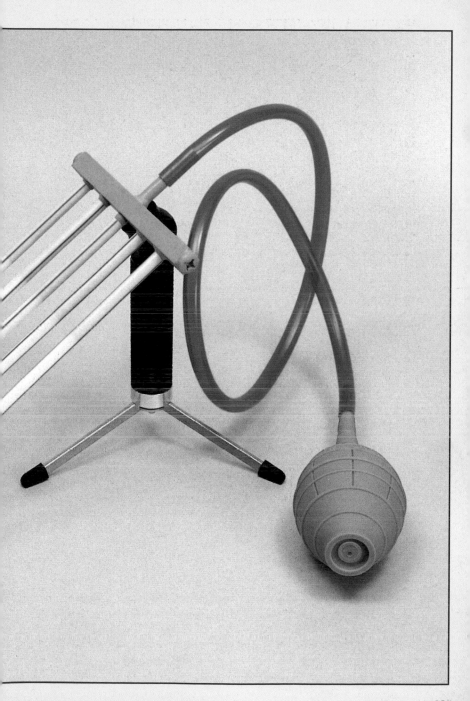

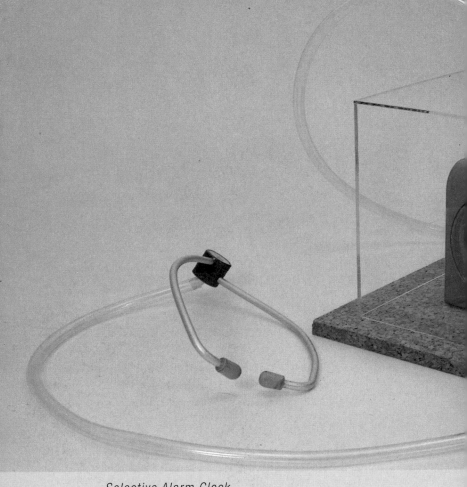

Selective Alarm Clock

✶ Only wakes the person who needs waking

The problem with alarm clocks is that they make too much noise. While that's fine for the person who wants to get up, it's not so great for those who'd like to keep on sleeping. It's especially difficult for large families and for husbands who want to sneak out and hit a few rounds of golf before their wives tell them that the car needs mending.

The Selective Alarm Clock is the solution. With its sound proof acrylic box and personal headphone attachment, this way the bell tolls only for the person who set the timer.

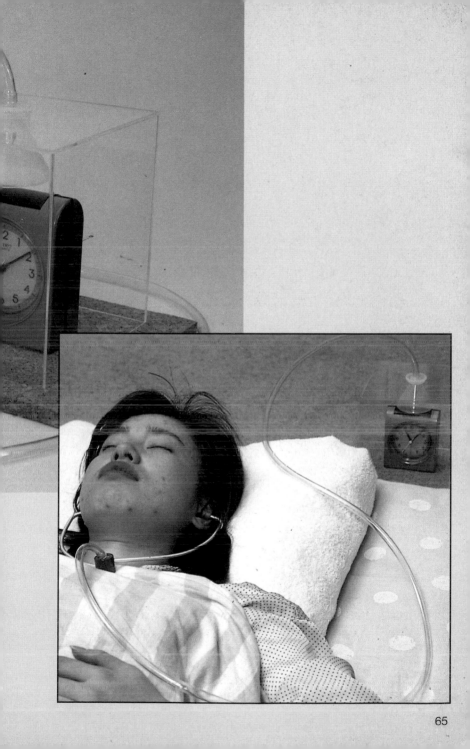

Portable Countryside

✳ If you can't take the country out of the boy, let him carry a bit around town

Big cities mean intimidating, towering skyscrapers and a puzzling maze of subways. Worst of all is the lack of natural sights and smells. Armed with a personal Portable Countryside the simple farm boy or born again naturalist need never feel too urban when weaving their way through the concrete jungle. From now on, the walk to work or school will always be pleasantly scenic and familiar.

We recommend using the backstreets whenever possible. Exposure to exhaust fumes will kill your plants.

Commuter's Fast Pass
✴ *Flash with panache*

Low many times have your fumbling fingers struggled to take off your gloves, fish out your pass and put it away again on cold mornings? And, sometimes, when you can't find it, the people behind you grow impatient. Frankly, it's embarrassing.

Consider the Commuter's Fast Pass for a quick in and out. For your convenience and comfort a see-through pocket for your commuting documents is attached to the outside of a warm woollen mitten. There's also a fingerless knit version for summer. An affected wave of the hand gets you through all the checkpoints and gives a certain air of chic to boot.

Since people on trains tend to stare at things, make sure your travel-pass photo is a good one.

珍道具

Traveller's Ice Pack

✳ The next best thing to staying at home in bed

When you have a fever or you're suffering from the world's worst hangover but you have to go out, wouldn't it be convenient if you could keep bundled up with a cold compress on your head until you get home again?

The Traveller's Ice Pack keeps your fever down while you do the town. It's slim, so you can walk down busy streets, and adjustable, so you can use it while riding trains or sitting at your desk. And if you're really ill, nothing beats it for making the trip to the chemist bearable.

The best part is that it's extremely noticeable: just what you need to illicit tons of sympathy from your friends and colleagues.

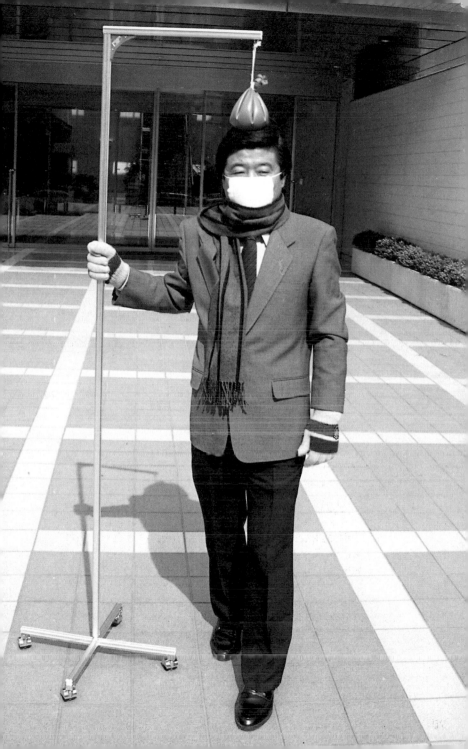

Cling Clang Cleaner

✳ Put your child's love of toys to a more practical use

E very parent who loves to sleep will attest unhappily to the fact that little children who have just learned how to walk love to show off their new skill. They 'cling clang' with their toys as soon as the sun comes up.

But don't discourage youthful exuberance, channel it to good use. The Cling Clang Cleaner is a lovable child's toy with attachments that also make it a mop or broom. This way, even though your child wakes you from sound slumber, at least the house will be sparking. If it's still in one piece.

Self-airing Shoes

✳ Furthers the fight against foot odour

Athlete's foot thrives in dark wet places. Fight fungus by keeping your socks dry and your shoes ventilated with Self-airing Shoes. Offending germs can neither rest nor reproduce as every step pumps a stream of wind through the shoe. You'll smell better and enjoy the sensation of walking on air.

Look out for embarrassing noises.

Scrub Gloves
✴ *Twelve times the power to scour*

The removal of stubborn stains from dirty pots and pans requires sturdy gloves, several sponges, scouring brushes, and a lot of elbow grease. Why not keep all these elements together in one place, ready for action? Cut down on washing time and effort with Scrub Gloves. Your grip will be a little restricted, so take care when handling your clean crockery or you may find your gloves will also reduce the number of dishes you own.

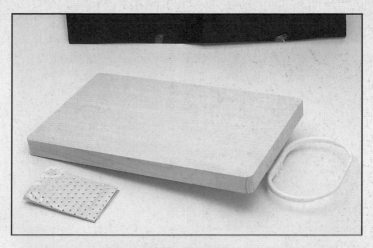

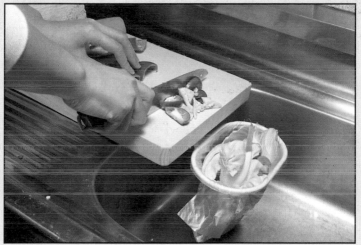

Chopping Block Litter Bag

✳ **Make your chopping block a trash-free zone**

Ｗhen you're chopping in the kitchen valuable space is taken up by rubbish that isn't destined to become dinner.

What you need is a rubbish bag right next to the cutting board. Bin those bones, sling those shells and veto that vegetable waste. Instant elimination means no clutter in the cutting area and no trash mixed back in with the food.

When the time comes to clear away the mess, throwing out the chopping refuse is a snap.

Baby Walking Slippers

✳ First steps need a helping hand...and two supporting feet

When babies learn to walk they spend a lot of time face down on the floor while their feet get used to the habit of walking. Baby Walking Slippers are designed to minimize the time and the stumbling required before your wee one gets his legs.

Velcro tabs attached to his little soles cling to the tops of the parent's slippers and, one-two, one-two, the child cuts through all those trial and error trip-u-lations and learns the technique straight from an expert.

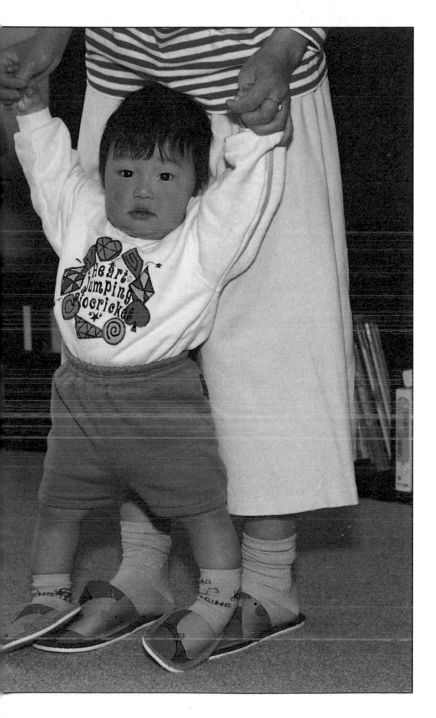

珍道具

Sit-up Table

✳ Take off calories while you take in calories

The most common excuse for not exercising is 'I can't find the time.' It's impossible to make that claim if you have a Sit-up Table. Now, while you're chewing you can also be crunching.

Beginners may want to start with just one sit-up every two to three bites, but advanced users will soon be able to squeeze off several exercises per mouthful.

If you're really lucky all that motion while you're trying to eat will result in a total loss of appetite. What could be better?

Shuffling/Scrubbing Shoes
✳ Tidy your tiles while you cut a rug

Most people like clean, but they don't like cleaning. Step into some Shuffling/Scrubbing Shoes and all you need do is stroll around your home and your floors will sparkle. Shuffle backwards, shuffle forwards, maybe even pretend you're skating. Then hose off the suds and enjoy a squeaky clean home.

You'll soon find that dancing is the best way to remove stubborn stains from the bathroom or kitchen and that your Shuffling/Scrubbing Shoes are the best way to get the job done. But exercise caution. It's hard enough to stand in these things, let alone disco. An unintentional spill in these could give break dancing a whole new meaning.

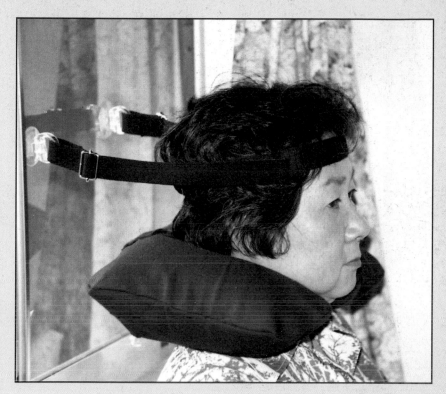

Good Posture Seat

✳ Straighten out your employees

Desks have a tendency to bring out the worst in a person's posture. When seated at their desks, some people constantly kick the legs of the table, some lean back like they're on holiday, and others tap their feet to music that's playing in their head. The Good Posture Seat puts a stop to all this fun and nonsense.

Sit down in one of these and you'll soon get the idea. With your spine pressed firmly into the back of the chair and your feet flat on the floor there'll be no more napping. Just imagine how much more productive you'll be.

Best of all, good habits die hard. Staff will often outlive their chairs.

Shiatsu Shoes

✳ **For adults who like massage and children who like to step on people**

In a household where everyone works hard, plays hard and comes home beat, it's difficult to find someone willing to massage your tired body. At the end of a long day only the kids have any energy left, but children don't know a thing about massage. Thanks to Shiatsu Shoes they don't have to.

Shiatsu Shoes have ten ringed plastic platforms on either sole. With a fifty-pound child, that means five pounds of direct pressure from each ring every time the small one lifts a foot. Since small children are notoriously unable to keep still, the rings will vibrate with every step, providing soothing relief for tired adults.

Bear in mind that your little masseuse may not have the cleanest shoes around. Always check their soles first or you may wind up with more on your back than you bargained for.

Portable Lamp Post
✳ *You'll never walk down a dark street again*

It's often not safe to be out at night, especially when you're a woman walking poorly lit streets. You know you shouldn't venture away from bright areas, but sometimes you just have to go down a dark road.

The Portable Lamp Post gives you an edge against potential attack from shadow-dwellers. It's bright and it's big and makes a loud clattering noise. If the light doesn't deter your attacker, the noise will.

Just two drawbacks. Immensely useful when the sun goes down, it's a pain to lug around in the daytime. And, hanging around under a lamp post may lead to you being mistaken for an enterprising prostitute.

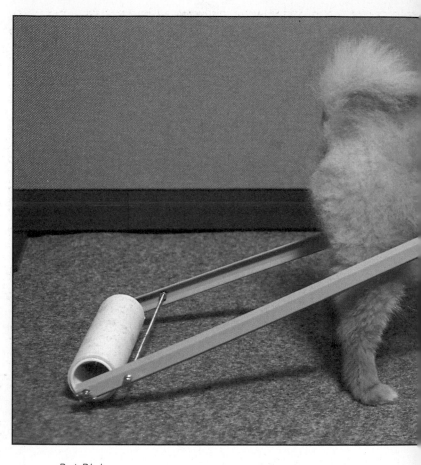

Pet Pick-up

✳ It's time dogs and cats cleaned up after themselves

Small animals are cute, but they're also messy. Apart from the smell and the noise there's also all that hair they leave behind them.

Hence the need for the Pet Pick-up. It lets the offending animal clear up its own debris almost in the same instant it's deposited. As long as the owner remembers to tear off the soiled outer sheet every now and then, the pet leaves a clean trail wherever it goes — with the added bonus of picking up human bits of trash too.

Your pets won't like it much at first, but they'll soon become as attached to it as they are to their own shadow.

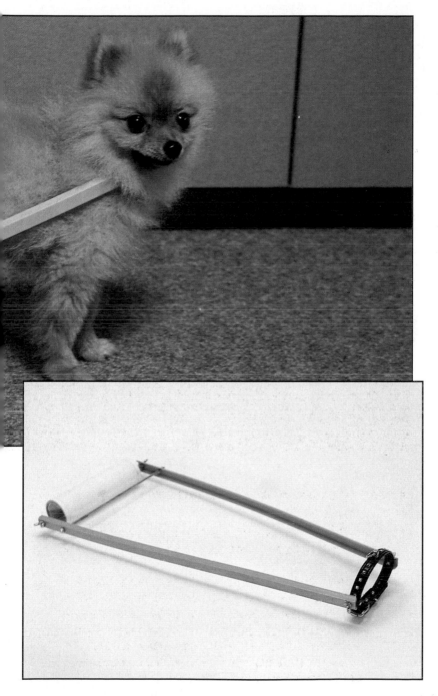

89

珍道具

Ladies' Leg Guard

✱ **Now fashion and foul weather go together**

Y̶ou've been splashed by passing vehicles one time too many. The Ladies' Leg Guard is designed especially for ladies who don't have the necessary dodging skills. They know best that stockings don't recover from muddy splatters the way an unclad leg responds to a quick wipe down. So, just attach one of these Leg Guards to the hem of your skirt and let the weather do its worst.

Self-Lighting Cigarettes
✳ *Neither wind nor rain will stop a smoke lighting again – they're match attached*

Now smokers needn't pat their pockets frantically, poke their heads into explosive gas ranges, or rub sticks together in a feeble attempt to get a light. With Self-Lighting Cigarettes the match is built in. This brilliant idea might also work for candles and fireworks, in fact anything that needs lighting.

The convenience far outweighs the discomfort of inhaling all the sulphur and gunpowder that's part of the match head.

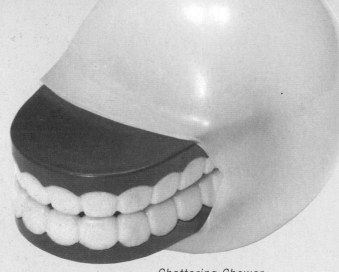

Chattering Chewer

✳ *Mashes up your meal so you don't have to*

Anthropologists tell us that our jaws are weaker today than they used to be. Tools like scissors and pliers have become substitutes for teeth, and the consumption of lots of processed food has decreased our power to devour.

One solution is to exercise the mouth more — one two, chew chew — but an easier alternative is just to go full force in the other direction and avoid anything that might give our infirm incisors trouble.

The Chattering Chewer accomplishes this nicely. By squeezing it a few times, the dentally deficient diner can crush up those hard bits before attempting to ingest them on his own. Great for people who have problems with nuts and bones.

No-slip Stair Scoops
✳ On rainy days, don't go anywhere without a pair

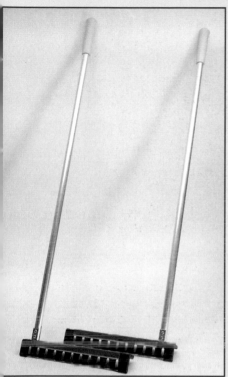

Even at the best of times it's pretty tough climbing those slippery marble or stone stairs that certain designers relish erecting. But try mounting them on a rainy day and it's next to impossible. Why flirt with disaster?

Stair Scoops are made with a no-slip rubbery edge, not just on the part that connects with your foot, but also on the bottom so they won't slide off the stair. Just put one down ahead of each step and you can ascend with mounting confidence.

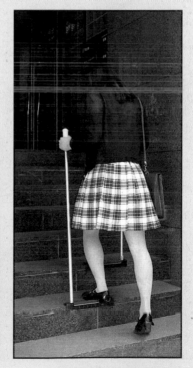

93

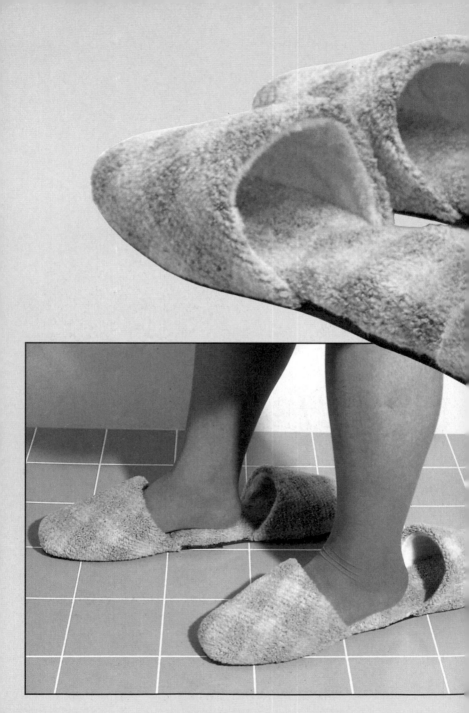

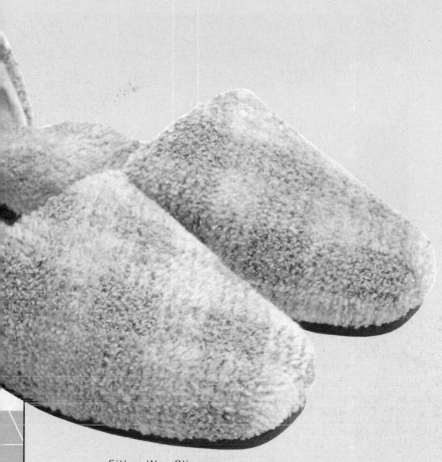

Either Way Slippers

✳ **Slippers that slip right on**

At the end of a long day, how good does it feel to slip into your slippers? At Japanese restaurants you can kick off your shoes and walk into a pair without even having to bend down: the hostess does the stooping and she makes sure the slippers are facing the right way. But at home there always seems to be a stubborn one that you have to kick around the room before you can get it facing the right way and get it on. Either Way Slippers put an end to this nuisance. No matter which way you step out of them when you're getting into your shoes in the morning, they'll always be facing in the right direction for you to step into them again when you get home.

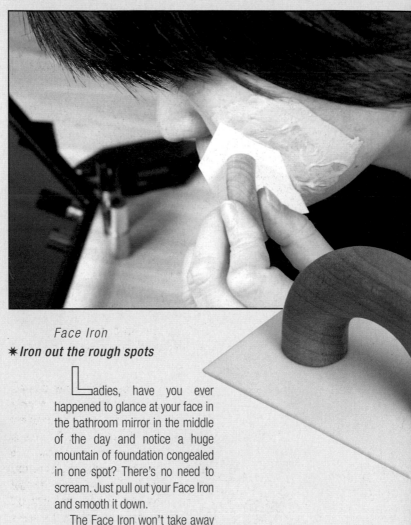

Face Iron
✳ Iron out the rough spots

Ladies, have you ever happened to glance at your face in the bathroom mirror in the middle of the day and notice a huge mountain of foundation congealed in one spot? There's no need to scream. Just pull out your Face Iron and smooth it down.

The Face Iron won't take away your wrinkles, but it will uncrease the make-up that you put over them. In the same way that a brick layer wipes away the excess cement, you can level off your face.

Later in the afternoon, watch yourself when you're tempted to open your mouth in a wide laugh. Your face will crack.

Lip Stamp
✳ *Pucker lips in a hurry*

Lipstick has its plusses, but it takes far too long to apply. You have to rub it on, smear it into place, wipe off the excess. Provided you have one that matches the size of your mouth, the Lip Stamp will get the stuff on more quickly and look just as good. It only has to take a second to achieve rich colourful lips.

The only problem is that it takes a little bit longer to get the colour onto the stamp – you have to rub it on, smear it into place, wipe off the excess...

Anywhere Office
✳ *Wherever you are, take your desk with you*

In a competitive working environment there are people who get in to the office early, leave late, and take lunch at their desks just to stay one step ahead of the crowd. With the Anywhere Office there's no longer a need to complain that there just aren't enough hours in the day. There are plenty – you just need to use them right.

Now you can run errands, go to meetings, take business trips and even go out to eat without ever having to leave your desk.

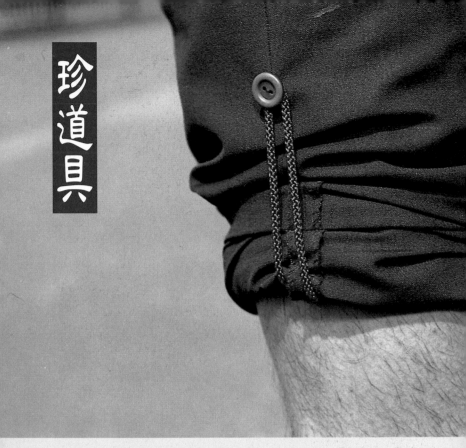

珍道具

Summer Suit Shorts

✳ *Being a businessman needn't mean being uncomfortable*

If you're a 'suit type' you'll find yourself staring longingly at kids in tank tops and cut-offs on long hot summer days, only able to console yourself and dismiss your sweaty discomfort with thoughts of your hefty pay cheque. When it comes to comfort, it seems the have-nots often have much more.

But now you can have the best of both worlds. By slipping into a pair of these trousers you can be as cool and comfortable as the beach bums. Fold them down just before you enter the office and your colleagues will be none the wiser as to your personal air conditioning.

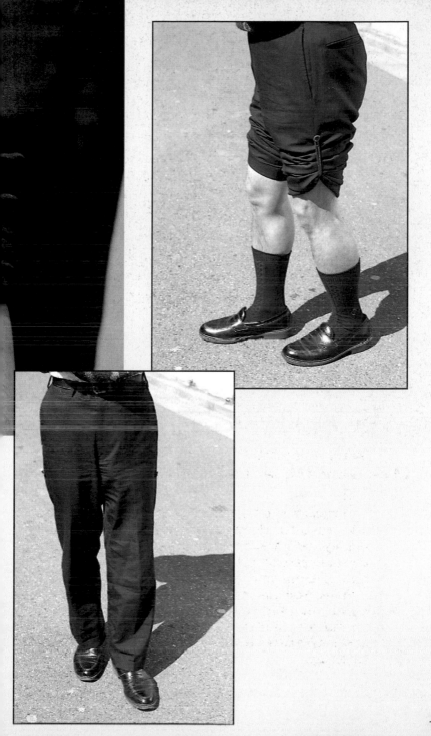

Rear View Camera

*See ahead of you with one eye, behind you with the other

When you look through your viewfinder to compose the perfect picture postcard, you see the world as a flat entity, already developed, trimmed and pasted into your photo album. Unfortunately, the people around you don't share in your freeze-frame detachment from reality. There's always the danger that the last picture you ever take will be of your subject screaming you a warning just before you're flattened by a bus.

The periscope on this Rear View Camera eliminates this fear. No more turning away in the middle of your camera adjustments to verify that there aren't any approaching hazards. A quick glance with your rear eye is all it takes.

Hold-It Helmet
✳Keep two eyes on your valuables, twenty-four hours a day

Big, bulky pieces of hard plastic that protect the head are very useful, but they leave tons of unexploited space. Why settle for just a helmet when it could do so much more?

With a metallic arm and clamp attached out in front you'll discover just how useful a helmet can be. It can hold your book on a crowded train, hold your pass when you're going through the station exit, even hold your camera when you want to take a picture. Helmet-top technology is still in its infancy. We've scarcely scratched the surface.

Self-Portrait Camera Stick
Do it yourself without the palaver

If you're travelling alone or as a couple, it's hard to get pictures with you in them. It can be embarrassing to have to ask someone to take the photo for you, confusing if they don't speak your language, even costly if the third party regards your camera as a gift!

With a fifty-seven-centimetre telescoping pole your dilemma is over. Expanding to three times its length for a full shot of you, your companion and your environs, your only problem will be that all your shots will capture you in the act of holding the pole. This could become a tiresome feature of your photo album – unless you really like poles.

Cigarette Joints

✳ ***Now chain smokers can live up to their name***

Don't get the wrong idea about our Cigarette Joints. They're not a new type of herbal cigarette. They're joining rings to make the old ones last longer.

When smokers gripe that their cigarette breaks aren't long enough, maybe what they really need are longer cigarettes. Cut the filters off a few extra smokes, use your Cigarette Joints to attach them to the one in your mouth, and you'll have a super-sized cigarette guaranteed to fulfil even your worst cravings.

Magnifying Fork

✷ *When it looks more, you'll eat less*

If the remaining sliver of cake is too small or the portions allowed by your diet too intolerably tiny, use a Magnifying Fork or the companion Expander Spoon and see more, sense more, taste more.

Tests show that nibbles eaten with the Magnifying Fork can be more satisfying than those had without, but much depends on one's state of mind. Sceptics tend to leave the table frustrated and unfulfilled. Gullible folk swear by it.

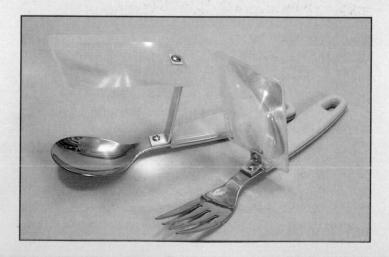

Swiss Army Gloves

✳ *Multi-purpose mittens that keep you ready for anything*

Those clever Swiss Army officers put a compass, screwdriver and nail clippers onto a pocket knife and revolutionized the average blade. Not to be caught bringing up the rear, the Americans have done similar things with flashlights that double as radios, and the Japanese have worked wonders with watches and calculators.

But it took the Art of Chindogu to realize that the tools very few humans ever hazard to leave behind are their hands.

Swiss Army Gloves provide a lighter, fork, corkscrew, bottle opener, pen, cutter, hole puncher and two screwdrivers all at the tips of your fingers — literally.

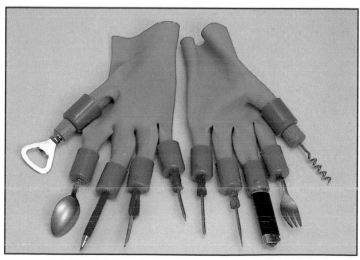

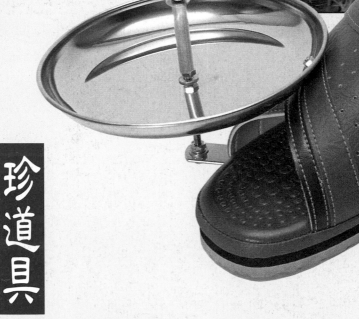

Skeeter Slippers

✳ Knock 'em dead wherever you tread

Those light-it-and-leave-it mosquito repellents work fine in the home. They smell nice and keep the pests away. But that won't keep the little blighters at bay if you have to go outside.

Try Skeeter Slippers. There's nothing to carry, nothing to spray and you wear the insect coils at foot level so that the rising smoke covers your entire body. Walk confidently through grass and next to streams: the mosquitoes simply won't bother you. And you'll be popular wherever you go because you'll carry your bug free environment with you.

珍道具

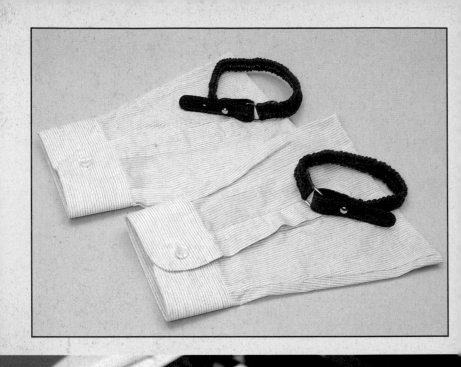

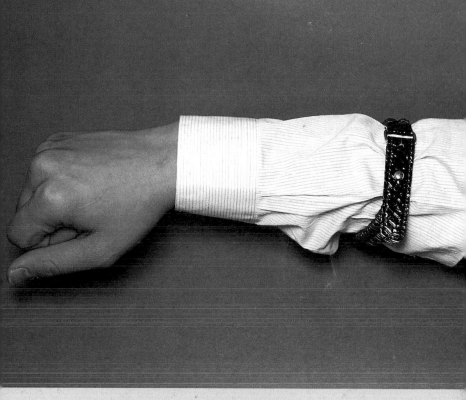

Shirtless Sleeves
✳ *A change of shirt you can keep in your pocket*

The downside of being an extremely successful business-person is the huge cleaning bill. Every day you have to put on a freshly-laundered shirt just so it can peek out from the sleeves of your jacket. Your boss notices very little apart from your back and hands at the keyboard, so why go to so much trouble? All that cleaning seems like a waste of time and money.

Why not try some Shirtless Sleeves? A good set of twenty or so can be stored in a briefcase and easily washed out in the sink. Clean sleeves every day and no one but you need know how long it's been since you last washed your shirt.

The Walls Have Eyes

✳ *Perform under scrutiny – somebody's watching you*

There's only one good substitute for self-discipline and that's the discipline inflicted by others. Too often students spend the whole day dreading the work ahead of them, then, when the rest of the world has gone to bed, they fidget, drift and eventually fall asleep.

That's why this screen of a hundred eyes is so useful. Pull it down when you finally want to get down to business and the power of these relentless stares and glares will keep you alert and concentrated.

Recommended only for study areas, though. This screen can be unnerving when placed in the bathroom or bedroom.

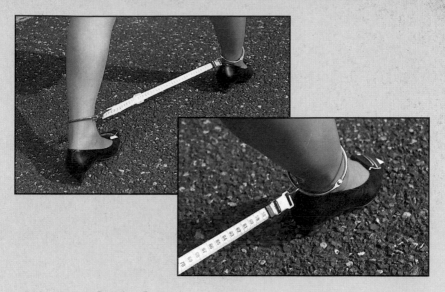

Precise Pacer

✳ *Accurate measurements that others can duplicate*

Counting off paces to measure distances works up to a point. But when those same steps are paced off again, there's often a different result – even when the same person does the pacing. The length of space between steps always varies and is hard to control.

That's why the Precise Pacer is so convenient. Strap one of these to your ankles and every step you take will measure the same. Forget centimetres, metres and kilometres, with the Precise Pacer you can make accurate assessments of your own.

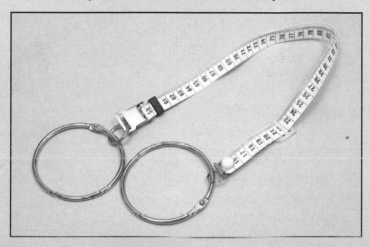

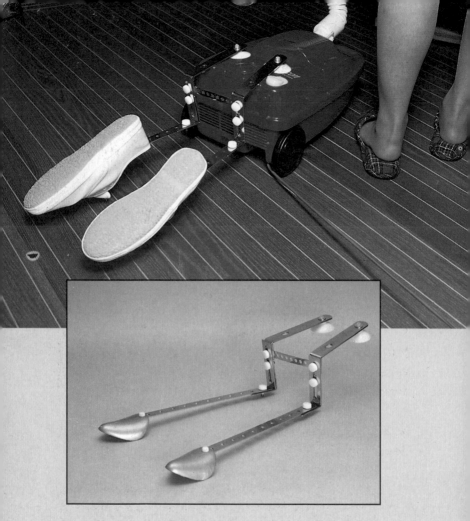

Vacuum Ventilator

✳ **Hoover the house and freshen your footwear...at the same time**

Vacuum cleaners don't just suck, they blow. A steady stream of clean filtered air, warmed by the motor, is expelled from the rear and escapes to the ceiling. Harness this resource and use it to air your tennis shoes.

Shoes moist with perspiration or the remains of a rainy day should be placed onto the Vacuum Ventilator and by the time the floors are clean, your shoes will be fresh, dry and ready to wear.

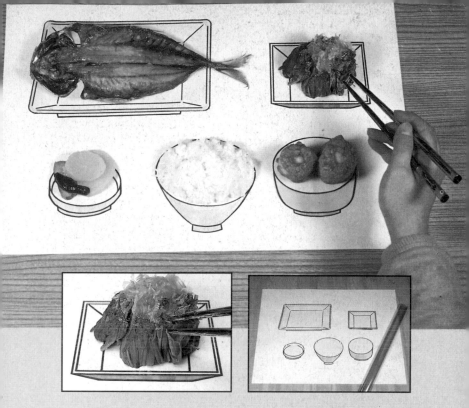

Plateless Placemats

✳ Enjoy the sight of a table full of dishes, but avoid the washing up

How often have you laid out cookies, rolls, sandwiches, nuts and hard boiled eggs on an array of pretty plates and dainty dishes only to wind up washing away three crumbs and an oil spot at the end of your meal and wondering why you bothered?

More than once? Then try Plateless Placemats. While permitting the elegant presentation of food, they save time, trouble and resources when it comes to cleaning up. With a single wipe or shake they are as good as new and, since they're double-sided, every now and then you can simply flip them over and treat yourself to the good china.

It's worth remembering that if you're having soup or stew for dinner, you'll need to make other arrangements.

Umbrella Brush

✳ When washing the ceiling doesn't mean washing your hair

Bathrooms are a breeding ground for mould and fungi. To make them easier to clean, a scrubbing brush attached to a hose means that a steady stream of water will wash away all the nasty stuff once you've loosened it from the tiles.

But what about the ceiling? That gets dirty, too. But you don't want all that murky water raining down on your head. That's why we routed the stream through rain's greatest enemy. The umbrella.

Don't be superstitious about opening your brolly in the house, or you'll get very wet.

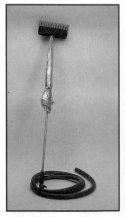

Stroll 'n' Style

* ***Strut your stuff while you do your do***

If *Saturday Night Fever*'s Tony Monero had had one of these, the movie would have been ten minutes shorter. With the Stroll 'n' Style, you don't need to spend precious time in front of the mirror with a hair drier every morning. Just slip this contraption on and off you go to work.

Every step you take will send a jet of air through your freshly washed head of hair; drying, styling, but requiring no extra time, energy or electricity.

Great for people with commutes that include long stretches of walking. Not as effective if you drive or take the bus.

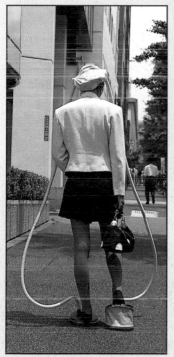

Deep Fry Spatter Shield

✳ Cook the goods and keep your distance

The taste, the sound, the smell of fried food is wonderful — everything's great, except the feel of it. If you're hit in the arm with a splash of oil, the pain will usually last longer than your dinner.

The Deep Fry Spatter Shield welcomes you to the new world of deep frying. Behind the safety of its clear plastic, you can not only see what you're doing but you're protected from those painful pops when the bits hit the pan. There's even a pair of chopsticks attached so you can load another piece of food in safety and with ease. If you're brave enough, you can even stick your face right in front of it and enjoy watching the droplets accumulate harmlessly on the other side.

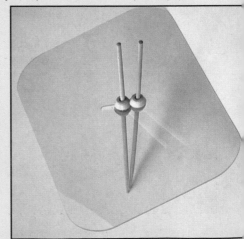

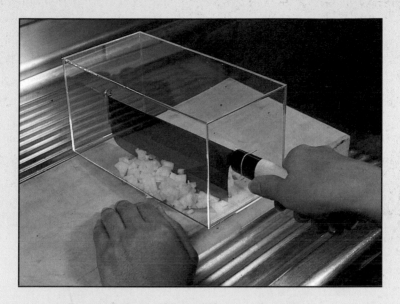

Tear-free Onion Chopper

✳ Cutting onions is no longer something to cry about

Chopping onions while holding back the tears is an age-old problem which is long overdue a remedy. Tissues are the obvious answer but, prevention being better than cure, an acrylic box with a big knife inside gets our vote.

The Tear-free Onion Chopper allows you to slice all you want without having to worry about eye-irritating odours and fumes. It's so safe, your child can help out in the kitchen and you won't need to fret over cut fingers.

It takes a little practice to cut an onion without holding it, but because you're not distracted by tears any more the job gets done in

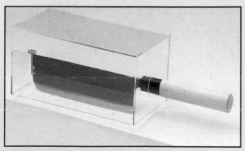

next to no time. And, ladies, the happy chopper now has more time to spend crying her eyes out in front of the afternoon soap operas.

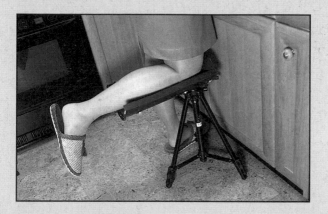

Leg-up Lounger
✳ Because leaning is easier than standing

Washing dishes, cooking, ironing – these are tedious and tiring chores that are hard to do sitting down. But with a Leg-up Lounger you no longer have to be on your feet all day.

Like the crane or the ostrich, who alternate feet so they don't get too tired, busy people can relieve half their load by using this cushioned leg rest attached to a collapsible tripod. At the end of the day, you'll feel like you've been sitting down half the time.

Use two and you'll feel like you've been sitting down all the time, but that would be greedy and just a little unstable.

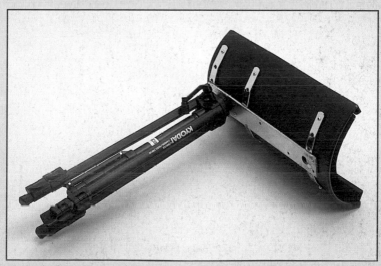

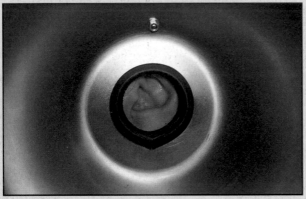

Ear Extender

✳ *Forget batteries and wires, hear a crisp new sound*

Who says that the humble, hand-crafted Art of Chindogu can't compete with the mass-produced technological wonders being churned out by those electronic manufacturing giants? The Ear Extender is a sturdily built, water-resistant, battery-free hearing aid that's virtually fool-proof and guaranteed to work whatever the circumstances. No transistors to fiddle with, no cells to charge, and you'll never have to worry about losing these down the drain!

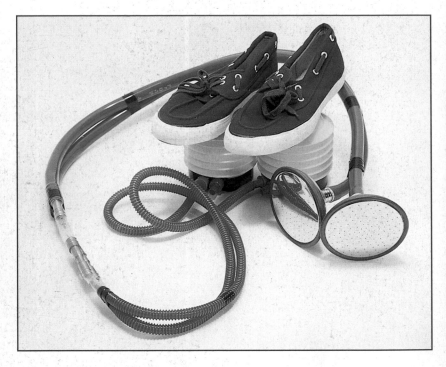

Bubble Bath Runner
✳ Soak your muscles in sweat, and then in the bath

Jogging on the spot keeps a body in shape, but what about all that energy exerted in the process? It usually goes to waste. This innovative gadget gives new meaning to the term 'running a bath'. Sprinkle in a little bath foam, slip into the Bubble Bath Runner shoes and you can create a cosy, sudsy paradise for escaping into once you've worked up a healthy sweat.

Alternatively, use it to give someone you love a whirlpool bath. And then it's your turn to relax in the tub while they get grimy again returning the favour.

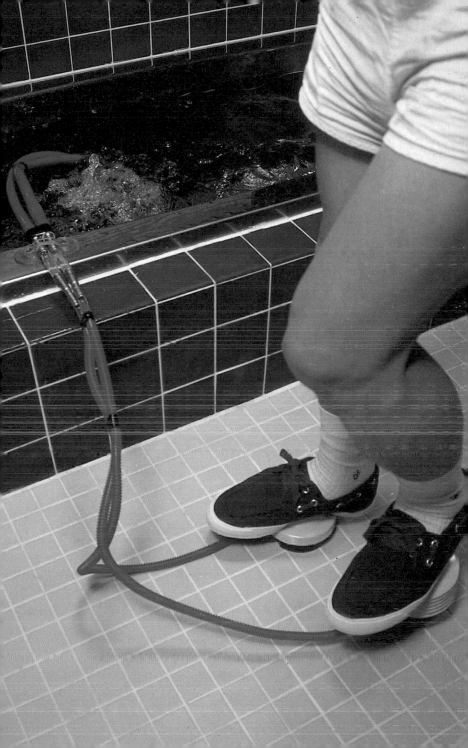

Desk Dreamer's Chin Rest

✳ *Eaten too much for lunch? Had one beer too many? Catch a quick nap at your desk*

Yes, we all know that, officially, you're not supposed to sleep in the office but you could argue that a cat nap will enhance your afternoon performance. How many working hours get wasted because people become jaded as the day wears on? How many employees would be saved from going home early because they feel 'out of it'? That's why this Chindogu for the chin is so handy. You don't get comfortable enough to snooze through the whole afternoon, but you do get a rest and without the stiff neck and ink smears on the forehead that can result from sleeping atop the desk proper.

And there's no need to fear about talking in your sleep – since the weight of the head rests on the jaw, your colleagues will enjoy perfect peace and quiet.

Infant Pleasing Camera
✳ *Makes all of baby's pictures pleasant*

Y̲ou can scream, wave your hands, jump up and down and call your child's name cajolingly, but all too often you'll still wind up with pictures of the poor kid crying. Change the theme of your photo album from frightened to fun with the Infant Pleasing Camera. And stop making a fool of yourself.

Just whip it out, blow on the mouthpiece and wiggle the box. Colourful snakes and streamers come to life to grab your child's attention while you grab a photo. They'll even look like they were enjoying the visit to the museum.

Scenic Extender

✳ *Zooming without the blooming expense*

The miracle of photography isn't in the camera, it's in the film. But that's not what the manufacturers would have you believe. They want you to think that to take good pictures you need costly peripherals like timers, zoom lenses and auto focusing. Well, we say 'ha' to that and have designed a gadget that echoes our sentiments.

The Scenic Extender will take you several metres closer to your subject without any expensive equipment. You can even use a disposable camera. Just squeeze the bulb and snap the picture.

Look out for our technological rivals' new top-of-the-range camera that flies, available soon in shops near you.

Baby Mops

✳ **Make your children work for their keep**

After the birth of a child there's always the temptation to say 'Yes, it's cute, but what can it do?' Until recently the answer was simply 'lie there and cry', but now babies can be put on the payroll, so to speak, almost as soon as they're born.

Just dress your young one in Baby Mops and set him or her down on any hard wood or tile floor that needs cleaning. You may at first need to get things started by calling to the infant from across the room, but pretty soon they'll be doing it all by themselves.

There's no child exploitation involved. The kid is doing what he does best anyway: crawling. But with Baby Mops he's also learning responsibility and a healthy work ethic.

TV for One
✳ Keep your television habits to yourself

A television can be an annoying presence to those who aren't watching it (or who are trying not to watch it). The light flickers, the movement distracts, the subjects on the screen vie with your needlepoint for your attention. You could invest in a new Personal-size TV, but they're about as much fun to watch as candlelight bouncing off a postage stamp.

Better by far is a TV for One. The unit – complete with twenty-one-inch screen – allows one person to enjoy their favourite show while another sleeps or does their homework. Adults can also enjoy sleazy night-time programmes while protecting the morals of young ones who aren't yet old enough to view said trash.

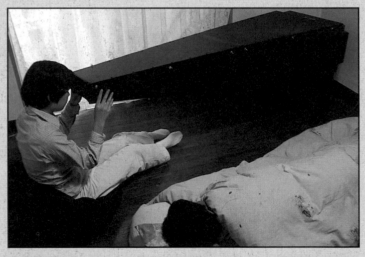

Umbrella Bell
✳ *Ring your thing to get past fast*

There's only one thing harder than moving through a crowd of sluggish Sunday strollers: moving through the same crowd on a rainy day. Umbrellas slow pedestrian traffic to a crawl.

Ever noticed how everyone jumps out of the way when a cyclist sounds his bell? That's the idea behind this umbrella attachment. Give it a ring whenever you need some space and by the time the others notice you're just another lowly stroller, they've already cleared the way.

Contact Lens Protectors
✳ *Never lose a lens again*

Exercise, wind, tears –
all these things can make a
contact lens pop out without
warning. When this happens
it's unlikely you'll ever find it: it's small, colourless, and you can only
see well out of one eye to look for it.

With the Contact Lens Protector you always know where the
missing lens is. In your Protector! No more hours wasted on your
hands and knees groping in vain. No more disquieting crunches
when someone approaches to ask what you're looking for. No more
eviction notices after you use your rent money to buy yet another
new lens.

Our Contact Lens Protectors keep lenses, lashes, and even
glass eyes in front of your face at all times.

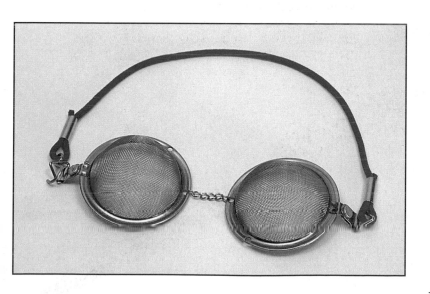

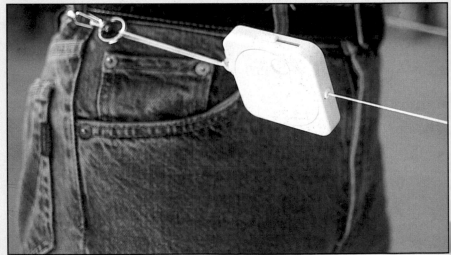

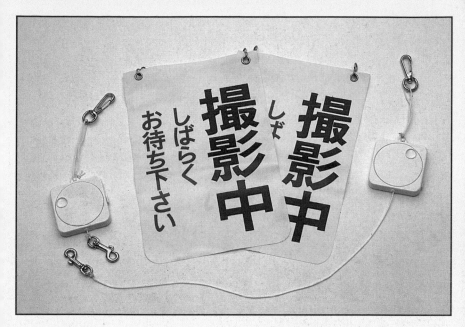

Picture Protector
✳ *Keeps creeps from sneaking into your shots*

Elvis Presley was notorious for picture crashing. He'd lean into a scene, unbeknown to the photographer or the subject, and when they got their film developed they'd be all shook up. But since it's unlikely that the King is available for more unscheduled photo shoots, the Picture Protector should be all you need.

With two cords that attach to the sides of the photographer and the subject an impassable fortress is created. For total idiots – who either don't see the string or think it's a finish line – each cord is also equipped with a sign that says 'Picture Taking in Progress'.

Mirror-bottom Wash Basin
✳ See your face as you're washing it

Sound obvious? Think about it. You don't really look at your face as you're washing, do you? Chances are you glance up at a mirror between washes. That may be the way it's always been, but it doesn't have to be this way in future.

Ladies especially can verify their make-up is coming off by watching as they're washing. The Mirror-bottom Wash Basin may well be the most effective face cleaning tool since the invention of water.

Drawbacks? Well, as you scrub, the water in the basin tends to cloud up, causing the user's face to fade out in front of their eyes. This can be alarming.

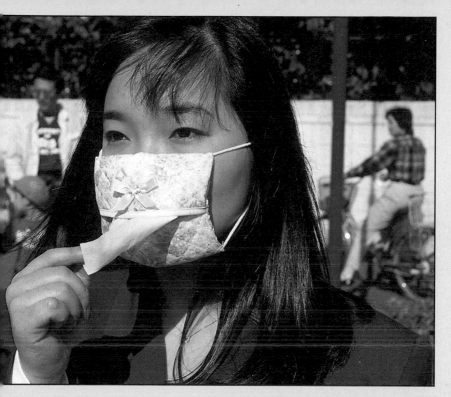

Tissue-equipped Face Mask
✳ A Kleenex container that's fast, fashionable, and right where you need it

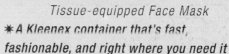

$\mathbb{S}$ome people wear masks when they suffer from hay fever. Put your mask to greater use by making it serve as an appropriately associated tissue holder too. Not only do Tissue-equipped Face Masks turn the pale white gauze numbers into something that keeps the wearer beautiful even though she's sick, they also add an extra barrier against the offending pollen: a large pile of tissues.

When the mask runs low on tissues, the wearer starts sniffling more and she knows it's time to top up her Kleenex.

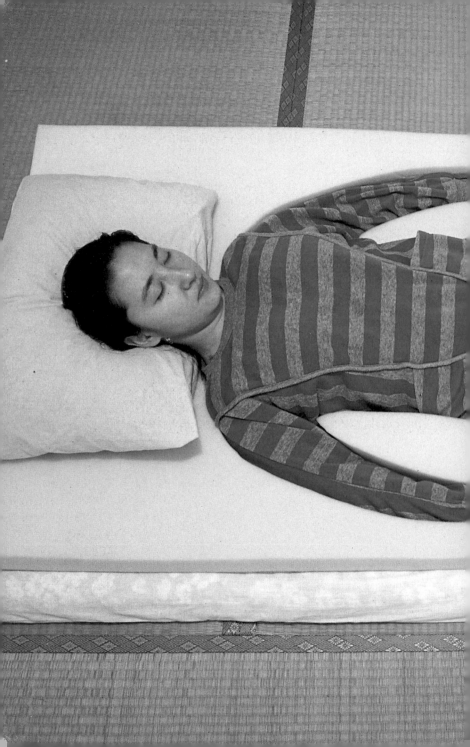

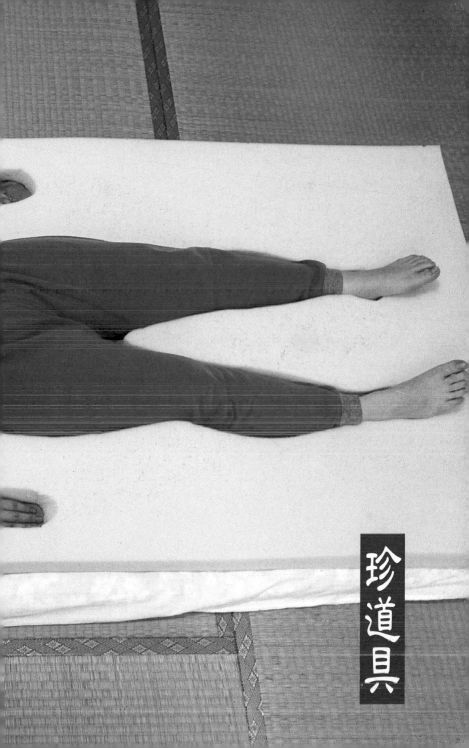

珍道具

Stay-in-place Sleeper

✳ The immobilizing mattress for middle of the night movers

Toss and turn when you're not sleeping well? Wiggling your covers and flinging your legs out of the bed are often manifestations of deeper troubles. The Stay-in-place Sleeper stops you from getting out of hand and from vacating your spot in the bed.

Slip into your niche when you lie down at night and the chances are you'll still be there in the morning. It's particularly useful for married couples who kick each other and wrestle over the sheets.

If you wake to find yourself on top of a pile of foam rubber — your Sleeper torn to shreds — this Chindogu probably isn't for you.

Umbrella Hat

✳Look! No hands!

People with packages, shoppers with shopping bags, or sightseers on strolls who just don't want to have to hold an umbrella up all the time: these are just a few of the folk who stand to benefit from the Umbrella Hat. No more accidentally jabbing people with your umbrella. And your arms are unencumbered to protect yourselves against jabs from others!

So it's a little less attractive when the sun comes out and you've got a closed umbrella sticking out of your head. But the merits of this handy headwear are otherwise endless.

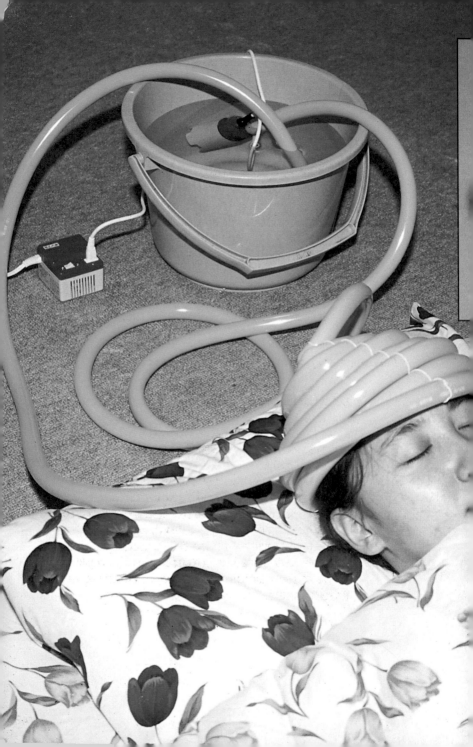

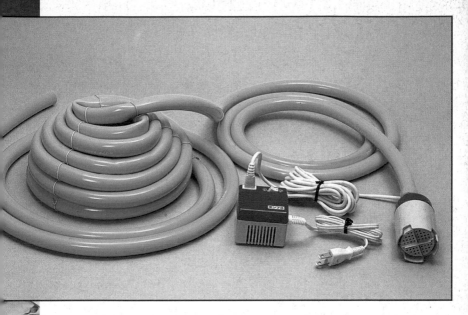

Auto-cycling Cool Compress

✳ Keep your fever down with a pampering pump

An ice pack or towel soaked in cold water is quite the best thing to bring a fever down. Within seconds these remedies start to cool your forehead, but said forehead also tends to heat up said instruments, necessitating frequent replacements. Blast those pesky laws of physics.

The Auto-cycling Cool Compress prevents you or your kindly caretaker from constantly having to run for more cold water. Just wrap the hose around your head and stick the pump in a bucket of water. As the cool liquid circulates, your fever is eased and the heat is carried back to the bucket. The sickly victim's benefits are endless: fever relief, a head and face massage, plus a low-pitched womb-like pulsing to lull him or her to sleep and aid fast recovery.

It's such a great feeling — watch out for hypochondriacs.

Environment Friendly Mosquito Nets
* The easy, economical and ecologically sound way to keep bugs at bay

In the summer months you might feel the need to erect huge tents of netting to keep the bugs out. You might even end up poisoning your own environment and polluting every breath of air just to kill a few tiny insects. Since the bugs are only going to bite the parts of your body they can get to, cover those and you've solved the problem!

The Environment Friendly Mosquito Net set comes with a face cover and bags for each hand and foot. Slip these on, make sure those flaps in your pyjamas are closed tightly, and you won't have a problem with the bugs the whole night.

Don't get a fright when you catch a glimpse of yourself in the bathroom mirror in the middle of the night! And when you wake up the next morning and shed the netting, watch out for some pretty hungry mosquitoes.

珍道具

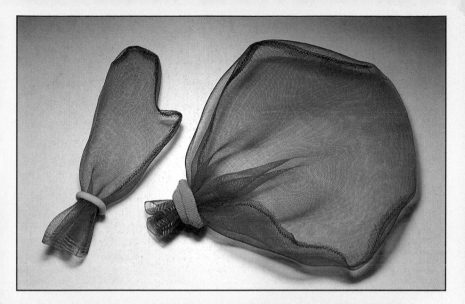

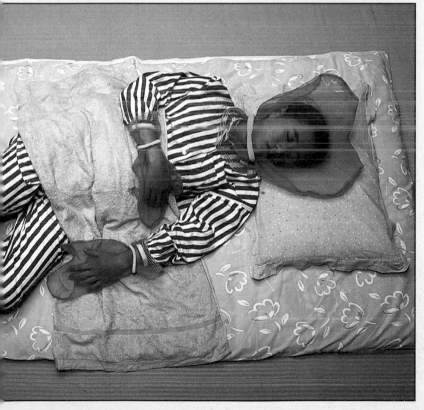

Portable Parking Spot

✳ Have space, will park

Victims of our socialization, a tiny white outline with a car inside is perceived to be a parking space. A large open area with no white line is not a parking space. The Portable Parking Spot allows you to turn this misconception to your own benefit. Simply roll the white painted strip around your car when you arrive and pick it up on your way out. Note: this last part tends to annoy other drivers if they're waiting to move in after you.

There's no need to worry about traffic wardens ticketing you. Suffering from the same disease of socialization, they seldom think to check the white line to verify that it really is a parking space and not simply a strip of paper that some clever person has brought with them.

Grin Grabber

✳ Forces even the crankiest folk to crack a smile

Doctors have noticed that people who smile frequently get fewer wrinkles and tend to have healthier skin. The irony is that by the time a doctor tells you this, you've already started to lose your looks and so have nothing to smile about.

The Grin Grabber solves this problem. Slip it on, stick the hooks into either side of your mouth, and pull the cord. It's guaranteed to correct a scowl. And it's economical, too. Just one person wearing this gadget is sure to get an entire room grinning madly.

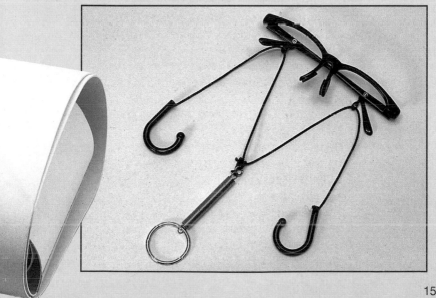

Drive-by Drier

✳ Air your attire while you tour the town

Now the natural feel and aroma of garments drip-dried in the great outdoors is easily achievable in the same time it takes to choose a drying cycle on your washing machine. You also get the luxury of choosing which section of town you want to smell like.

The Drive-by Drier renders a full load of laundry wearable in just twenty minutes on the highway, twenty-five in the city. (Estimates based on Tokyo test drives; actual drying time may vary.)

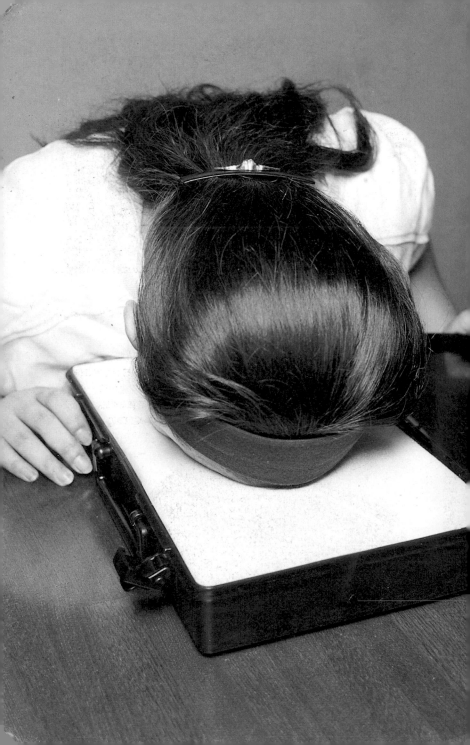

Geisha On The Go

✴ **No time left to put on your make-up? Take a bow and you're ready to rock**

The evolution of cosmetic particle application has finally taken an intelligent turn. The powder goes on at quantum speed — with one quick dip. And you don't even need a mirror; missing a spot with Geisha On The Go would be like going swimming and not getting wet.

Anyone desiring information on how to join the
International Chindogu Academy
may write to:

Kenji Kawakami
Chindogu Academy
5–12–12 Koishikawa
Bunkyo-Ku
Tokyo 113
Japan

or on e-mail at: http://www.chindogu.com